The Biology Series

CELL MACHINERY

by Robert Leftwich
Illustrated by Larry Weaver

Cognizant of the quality of illustrations characterizing the current biology textbooks, the author has developed this series with a tone of reinforcement and enrichment for troublesome or difficult concepts.

Acknowledgment: All the electron micrographs used are through the courtesy of Dr. David M. Phillips, Assistant Professor of Biology, Washington University, St. Louis. The fixative for the electron micrographs was gluteraldehydeosmium. The magnification is indicated in the teacher's guide section associated with each transparency.

CONTENTS

Page
* 1. Early Microscopes
* 2. Basic Microscopes
    2A. Basic Microscopes for Cell Study
* 3. Plant Cell
    3A. A Plant Cell
    3B. Basic Plant Tissue Type Cell
* 4. Animal Cell
    4A. An Animal Cell
    4B. Basic Animal Tissue Types
    4C. A Generalized Plant-Animal Cell
* 5. Nucleus
* 6. Chromosomes
    6A. Chromosome Replication
* 7. Cell Division
    7A. Mitosis
* 8. Centriole and Flagellum
    8A. Centriole and Flagellum
* 9. Mitochondria
    9A. Mitochondria
*10. Golgi Complex
    10A. Golgi Complex
*11. Endoplasmic Reticulum with Ribosomes
    11A. Endoplasmic Reticulum and Ribosomes
*12. Plasma Membrane
 13–16. Unit Test on the Cell

* Indicates full color transparency.

---

*Page 1 – EARLY MICROSCOPES

CONCEPT: The twentieth century concept of the cell reflects five centuries of instrument development, exacting technology, and pervasive synthesis.

BACKGROUND INFORMATION: The two most pervasive biological theories emerged almost simultaneously in a definite form during the 19th century. The theory of evolution was heralded with considerable ceremony and controversy. It reflected the work of two great minds — Charles Darwin and Russell Wallace. The cell theory generated less excitement and provoked almost no controversy. The rationale for the cell theory mirrored the painstaking work of several generations of scientists, technicians, and hobbyists. The evidence for a theory of evolution was macroscopic while the evidence for a cell theory was microscopic.

MILLIKEN PUBLISHING CO.

The mean diameter of a cell is 10 microns. Man's eye, however, could resolve objects with diameters of only about 100 microns. Therefore, before man could effectively study the cell, his power to resolve had to be improved by a factor of 10. It is not surprising that the problem would be solved by two spectacle makers, Jans and Zacharia Janssen. In 1590 they combined a convex objective lens with a concave eyepiece to produce the first compound microscope. There is little doubt that the resolution was poor.

A mathematician, Johannes Kepler (1571-1630), suggested but never developed an improved microscope using a convex lens for both the objective lens and the ocular lens. This was the basic microscope model that was to experience constant improvement. Robert Hooke (1635-1703) used an improved compound microscope and dark field technique to view a thin section of cork. Without comprehending the nature of what he saw, he unfortunately called the little box-like cavities he saw "cells" (little rooms). Corpuscles (little bodies) would have been a much more descriptive term for cells in general.

Anton Von Leeuwenhoek (1632-1723) mastered the art of grinding simple lenses of short focal length. He turned his microscope on a variety of things but created the greatest stir in the civilized world when he reported the presence of little beasts in a drop of water. Leeuwenhoek is accredited with initiating the microscopic study of living things.

During the next hundred years, techniques for the preparation of microscopic study material were worked out, and the microscope was applied to the study of tissues of various organisms. An inductive generalization began to emerge. In 1759 Casper Wolff pointed out the consistency in composition of globular units in embryological material. In 1802 M. de Mirbel expanded Wolff's generalization to include a cellular nature for plant tissues. The structural unity of the cell was becoming generally accepted. Historically, R. J. H. Dutrochet in 1824 advanced this inductive generalization to include all organisms. By 1831 Robert Brown had resolved that the cell had a nucleus. Credit for the pervasive synthesis of the main ideas of the cell theory traditionally has been given to Matthias Schleiden, 1838, a botanist, and Theodore Schwann, 1839, a zoologist. Certainly Schwann's and Schleiden's work served to advance the idea that the cell was the unit of organization and the unit of composition of all living things. Rudolf Virchow in 1858 supplemented the cell doctrine with his principle of biogenesis. He pointed out that all cells come from pre-existing cells.

Since then, the focus of interest in the cell has experienced great change. The cell is currently recognized as the working area for the complex organelles that abide within its membrane. This reflects the cell's role as a unity of physiological activity rather than the ultimate structural unit of living things.

FOR FURTHER STUDY: 1. Find out why Purkinje's term "protoplasm" is rapidly falling into disuse. 2. Discuss the implication of Virchow's declaration that all living things arise from pre-existing living things. (Where did the first living thing come from)? 3. Find out why some biologists argue that the single-celled organisms like paramecium should really be called acellular rather than unicellular. 4. Do viruses violate the cell theory? Qualify your answer.

*Page 2 – BASIC MICROSCOPES

CONCEPT: Man's continued penetration of the enigma of life as it was manifest in the corpuscular unit, the cell, has been tied to the development of instruments that have expanded his resolving power.

BACKGROUND INFORMATION: The past relationship between the growth of the corpuscular or cell theory and microscopic innovation is a fine example of the dependence of scientific advance on the advance of supporting instruments. The bright field microscope used in the early 19th century was mechanically improved but not significantly better optically than the "Kepler" type instrument used by Hooke. The instrument began to take it's final form when Charles Spencer (1813-81) introduced objective lenses with high numerical apertures (N.A.). Ernst Abbe (1840-1905) helped matters even more with the development of apochromatic objectives and the accompanying eyepiece which minimized the spherical and color aberration that had been found so distracting. In 1853 Francis Wenham improved the specimen's viewing contrast through the development of a condenser. The working distance between the specimen and the objective lens was minimized by R. B. Tolles in 1874 through the development of an immersion objective which unified the specimen and the objective lens. The coal tar dyes further accelerated cell study by providing additional specimen contrast through differential staining techniques. Microtomes were developed for the preparation of very thin specimens. The bright-field microscope that has emerged is currently capable of resolving objects about 200 angstrom units apart. The physical limitation of the instrument is about 2000X, but the level of high quality definition is only about half of that. Thus the practical working magnification of the bright-field microscope is only about 1000X.

Frits Zernike in 1935 gave birth to a new generation of microscopes. Visually, the phase contrast microscope is almost identical to the bright-field instrument. The phase contrast microscope is able to accent minute density differences in otherwise almost transparent specimens through a quarter wave shifting of undiffracted light waves passing through the specimen. This quarter wave shift, is typically accomplished through the use of an annular diaphragm usually placed in the focal plane of the condenser and the interpretation of the measure of interference is made by a phase-shifting element generally placed in the rear focal plane of the objective. The phase contrast microscope has proven particularly useful in the non-stain study of living transparent specimens.

The instrument which has revealed the most about the fine structures of the cell is the electron microscope. The resolving index of the electron microscope is about 10 times that of the bright field microscope. Objects only 20 angstrom units apart can be resolved by the electron microscope. A general understanding of the mechanics of the electron microscope may be had by comparing it to an inverted light microscope. The large cluster at the top of the device serves as a source of electrons, the illuminator. A magnetic condenser lens focuses a beam of electrons on the dead, fixed specimen held in an evacuated tube. As the condensed electrons move past the specimen, they may be absorbed, scattered, or allowed to pass unaffected. The emerging electrons are brought in to focus by means of a magnetic objective lens. The focused electrons are subjected to a magnetic projector lens. The magnification is varied by varying the current in the projector lens. The projected electrons are allowed to strike a fluorescent screen or a photographic plate. The image produced is in reality a shadow of the specimen wrought by the measure to which the electrons were scattered, extenuated, or unextenuated.

Thus, the electron microscope has expanded the eye's resolving power about 500 times — a long way from the resolving power afforded by the microscopes of Hooke and Leeuwenhoek.

Page 2A — Basic Microscopes for Cell Study

Have students complete this page according to transparency 2.

*Page 3 — PLANT CELL

CONCEPT: The rapid perfection of the electron microscope, improved cytochemical techniques, and the ability to separate the cellular organelles through homogenation and centrifugation provided a new horizon for the study of the ultrastructure of the cell.

THE SUBJECT: Root Meristematic cell of the broad bean *Vicia faba*, 9,000X.

BACKGROUND INFORMATION: It was the distinctiveness of the cell wall of the cork cells that drew Hooke's attention to the regularity of presentation of plant tissue. The presence of a cell wall has since been used to characterize plant cells. Before the 18th century ended, the microscopist had resolved green bodies in some plant cells along with the cytoplasm. The living plant cell was then a rigid wall confining cytoplasm and green bodies. Early in the 19th century Robert Brown added the nucleus and the nucleolus to the cellular repertoire. The development of dyes and their use to stain cells prepared for study added other organelles. The chromosomes were detected in dividing cells by the middle of the century. Camillo Golgi resolved a body he called the "internal reticular apparatus." Mitochondria showed up as rods or distinct filaments after treatments of Janus green B and neutral red dye. The stains also made it clear that the soft parenchymous tissues of the plant had a large vacuole which occupied the central part of the cell. The cytoplasm was crowded to the sides. The cytoplasm showed up as sometimes smooth and sometimes rather granular, but the endoplasmic reticulum was identified. Starch leucoplasts were identified. By the beginning of the 20th century most of the major land marks of the cell were recognized.

There was some understanding of the chemical character of some of the constituents of the cell by this time, but the real breakthrough came when methods were developed for the homogenization and separation of the cellular complements through the use of the centrifuge. When tissue was homogenized and centrifuged for 10 minutes at 600 g (units of gravitational force) the nucleus and any intact cells separated out. At 8000 g for 10 minutes the mitochondria separated out. At 25,000 g lysosomes separated out. At 105,000 g for 10 minutes the microsomes and endoplasmic reticulum separated out.

Thus it became possible to separate, maintain, and study the cellular organelles. Using this technique, in addition to refinement in knowledge of biochemistry as applied to the cell, the chemical nature and some of the chemical activity engaged in was learned before the super structure was resolved. By the 1950's the electron microscope was turned on cellular organelles. The chloroplast and mitochondria were shown to have elaborate internal structures. The membrane was shown to be trilaminar. The cell wall consisted of long cellulose fibers. Cells were connected by desmosomes. The cytoplast was shown to consist of tube-like and passage-like areas called the endoplasmic reticulum. The endoplasmic reticulum was frequently lined with small spherical structures called microsomes known to be associated with protein synthesis. The Golgi's disputed organelle was positively affirmed and called Golgi complex. Lysosomes were identified.

The current concept of the cell, the cell of the electron microscopist, the cell of the biochemist is a far cry from Hooke's cell. A cell is a working area for a complex of organelles and macromolecules related and directed by chemical reactions.

FOR FURTHER STUDY: 1. Find out what use the plant cell makes of the cell wall and what limitation the cell wall places on a cell. 2. Find out what holds plant cells together. 3. What is misleading about a generalized plant cell presentation? 4. What is instructive about a generalized cell presentation?

Page 3A — A Plant Cell

Have students complete this page, using transparency 3 as a guide.

Page 3B — Basic Plant Tissue Type Cell

Students are to complete this page.
- A. — vascular
- B. — sclerenchymous
- C. — collenchymous
- D. — sclerenchymous
- E. — parenchymous
- F. — vascular
- G. — sclerenchymous

*Page 4 — ANIMAL CELL

CONCEPT: There is unity in cells despite the diversity they manifest.

THE SUBJECT: Tissue culture cell, fibroblast, connective tissue cell of a Chinese hamster 12,000X.

BIOLOGY — Cell Machinery

BACKGROUND INFORMATION: The French biologist, Andre Lwoff, points out that the cellular unit is unified by the unity of plan (nucleus imbedded in cytoplasm), unity of function (uniform metabolism), and unity of composition (consistence of macromolecular make-up). But cells are different and diverse. Animal and plant cells are frequently heralded as reflecting the extremes of cellular diversity. Yet a study of animal tissue or plant tissue will show more diversity between the cells of different tissues of the same organism than exist between cells of analagous tissues of organisms as different as plants and animals.

The mean diameter of a cell is 10 microns, but this tells very little about cells since the range in diameter is from about 2 microns to the size of an ostrich egg. The cortical cells are tetrakaidecahedron (14 sides) but the nerve cell process is a skinny cylinder. It usually comes as a surprise to students to learn that typical cells are not brick or sphere shaped.

Many are equally surprised to learn that generalized plant and generalized animal cells are synthesis products and not real cells. There are however, type cells which characterize the basic tissue types. Plant cells may be catalogued as meristematic, parenchymous, collenchymous, sclerenchymous or vascular. Animal tisues may be classified as epithelial, supporting, muscle, nerve, or vascular.

There are certain ubiquitous organelles. Ribosomes have been found in all kinds of cells, endoplasmic reticulum appears regularly, Golgi complexes are widespread, mitochondria are almost universal, the nucleus and the distinctive nuceolus are omnipresent, and finally the trilaminar membrane becomes monotonous. Some cellular organelles are characteristic of large phylogenetic groups. The presence of a cell wall is distinctive for the higher plants. However, chloroplasts are not characteristic of plant cells in general, but of parenchymous mesenchyme light-exposed cells only. But chloroplasts are conspicuously absent from animal cells of all types as are centrioles from the cells of higher plants. Lysosomes have not been demonstrated in plant cells despite the fact that there is a need for the intracellular digestive and lysis action they are known to promote.

It would be anthropomorphizing to say cells have personality, but it is true that cells do display a high level of diversity.

FOR FURTHER STUDY: 1. Centrioles are found in the cells of some fungi, some algae, and the higher animals. Would this unify these taxonomic groups? 2. Would the absence of centrioles in vascular plants disassociate the vascular plants with the algae that have centrioles?

Page 4A – An Animal Cell

Have students complete this page according to transparency 4.

Page 4B – Basic Animal Tissue Types

Students are to complete this page.

    A. – epithelial

    B. – vascular (blood)

    C. – epithelial

    D. – connective

    E. – supporting

    F. – muscle

    G. – nerve

Page 4C – A Generalized Plant-Animal Cell

Students are to complete this page using the following labels:

    A. – an animal cell

    B. – a plant cell

    C. – plasma membrane

    D. – lysosome

    E. – endoplasmic reticulum

    F. – vesicle

    G. – nucleolus

    H. – vacuole

    I. – cytoplasm

    J. – a chloroplast

    K. – centriole

    L. – nucleus

    M.– Golgi

*Page 5 – NUCLEUS

CONCEPT: The nucleus, the information center, is usually the largest cellular organelle. It is chemically distinguished as the almost exclusive reservoir of deoxyribonucleic acid, the staple of genetics.

THE SUBJECT: Root meristematic cells of the broad bean *Vicia faba*, 6500X.

BACKGROUND INFORMATION: Robert Browne is credited with the detection of the first cell organelle. In 1833, while wor' g with tissue from an orchid, he noted the constant appearance of a large, dense body. Since then the nucleus has demanded and received more study and analysis than all the other organelles combined. It is a constant and essential cellular component, present at some time during the life of a cell in an organized, membrane-enveloped form in all organisms except those of Kingdom Monera. The Monerans have nuclear material which distinctly lacks an encasing membrane.

The nucleus is usually round or ovoid in form and frequently eccentrically positioned within the cell. When the use of dyes became widespread, the nucleus could be distinguished by its response to basophylic dye. The densely stained network of the vegetative cell nucleus was thus given the name chromatin. One or more areas in the nucleus of the vegetative cell would characteristically appear as dense, dark bodies and were called nucleoli.

The events associated with nuclear division were observed and described before the end of the 19th century. The 20th century started with the establishment of an association between genetic factors and chromosome behavior. This intensified interest in the nucleus.

BIOLOGY – Cell Machinery

Cytochemical techniques were developed which unravelled some of the chemistry of the nucleus. The Feulgen reaction made it clear that the nucleolus and the chromatin of the nucleus were chemically different. The nucleolus was Feulgen negative and, on analysis, was found to consist of ribonucleic acid and protein. The nucleolus waned at the inception of cell division and reappeared following cleavage. The nucleolus is morphologically distinguished by the presence of fibers.

Chemical analysis of the chromatin showed it to consist of deosyribonucleic acid and a protein complexer, histone. During cell division the chromatin gets organized into dense thread-like bodies called chromosomes. The shape or appearance of a chromosome depends on the locus of the centromere. The number of chromosomes in a cell is a function of the species represented and of the ploid level reflected.

The enveloping membrane is very flexible and consists of two membrane units. The membrane unit itself consists of three basic layers. Some specimens show nuclear pores with septate diaphragms capable of opening or closing. Within the nucleus there are small bodies of the size, shape, and chemical composition of cytoplasmic ribosomes. The meaning of this is not clear. The possibility that nuclear ribosomes might synthesize the nuclear protein histone is, of course, an attractive one. There is speculation that perhaps all ribosomes are synthesized in the nucleus.

The nucleus is sufficiently large to facilitate cell surgery. Nuclei can be removed or exchanged between cells using microsurgical techniques. The removal of the nucleus leads to the death of the cell. A nuclear exchange between cells means that the new nucleus will eventually take over and direct the new cell along lines consistent with the activity the nucleus directed in its original donor cell.

It is widely accepted that the genetic specifications of all cells in an organism are the same. The nuclei of cells of different tissues then differ mainly in respect to what part of their genetic instruction repertoire they use. The mechanism for such state of affairs has proven tremendously elusive.

The nucleus then has an ultra-structure: chromatin or chromosomes depending on metabolic state, nuclear ribosomes (?), nucleoli with fibers, nuclear lymph, and a not so elastic double membrane housing. The nucleus is the innovator and director of the chemistry of the cell.

FOR FURTHER STUDY: 1. Find out what happens to the nucleus of the mammalian red blood cell. 2. Discuss the significance of the absence of a nuclear membrane in the most primitive kingdom of organisms, the Monera. 3. Find out about the protozoa with vegetative and reproductive nuclei.

*Page 6 – CHROMOSOMES

CONCEPT: Considerable evidence exists which suggests that the Feulgen positive thread-like bodies visible in the nucleus during karyokinesis are indeed the bearers of the heriditary unit — the gene.

THE SUBJECT: A metaphase plate polar view of a fungus gnat *(Sciara)* spermatid, 35,000X.

BACKGROUND INFORMATION: Eduard Strasburger (1844-1912), a German botanist, in 1875 published the first description of the dark colored bodies visible only during nuclear division. The German cytologist, Walther Flemming, detected the longitudinal splitting of these colored bodies (chromosomes) in 1882. This clarified the process of chromosome cleavage during cell division. A marriage of cytology and genetics was affected by the American, Walter Sutton, in 1902 with his correlation between the chromosome reduction division he saw in the grasshopper spermatogonia and the factor segregation associated with Mendelian hereditary factors. By 1927 the American, H. J. Muller, had discovered a correlation between X-ray induced chromosome irregularities and phenotypic nutations.

The way was paved for the locus concept of the gene. Genes were arranged on the chromosome like beads of a necklace. The frequency with which genes on the same chromosome recombined with a different set of genes was used to make maps of the chromosomes. Visual support for the string of beads concept was provided by the very abnormal giant chromosomes of the fruit fly salivary gland. These chromosomes showed alternating bands of dark and light material.

Chemical analysis shaped the next concept about the chromosome. It was discovered that the ratio of the nucleotides adenine to thymine, and cytosine to guanine, was always 1:1, independent of the species studied. The ratio of adenine: cytosine or of thymine: guanine varied from species to species. The next contribution was made by a physicist. X-ray diffraction study of DNA showed markings which could best be interpreted as giant spirals which repeated themselves at regular intervals.

The synthesis was performed by James Watson, an American, and Francis Crick, a Britisher. They proposed that the DNA molecule was a spiraled double helix. The two spirals were held together by specific cross bonding between adenine and thymine or cytosine and guanine.

The Watson-Crick model of DNA has been so widely accepted and so well publicized that the inherent danger lies in forgetting that it is a theory, not a fact. Although the double thread construction is visible in certain bright field microscope preparations, the electron microscope resolution has provided no illumination at all and has certainly been a major disappointment to cytologists. That is, electron micrographs give no hint of a spiraled double helix.

Thus the chromosome is visually a coiled, twisted, Feulgen positive complex of DNA and histone. It has a centromere (a handle). It does bear the genes, but whether they dwell at a particular spot, are a finite length, are variable, are expressed in a specific number of nucleotides, or are the smallest genetic unit capable of producing a change if chemical variation occurs are matters awaiting further clarification.

FOR FURTHER STUDY: 1. Genes are variously defined as recons, cistrons, or mutons. Find out how these concepts differ and why each is useful. 2. Find out the number of chromosomes typical of several different organisms. Discuss the significance or lack of significance of the similarities or differences.

BIOLOGY – Cell Machinery

Page 6A — Chromosome Replication

Have students complete this page.

    A. — DNA molecular mock-up
    B. — DNA double helix
    C. — DNA unzipping

*Page 7 — CELL DIVISION

CONCEPT: No biological phenomenon has been more thoroughly studied, more precisely described, yet more poorly understood than that of nuclear division, karyokinesis, and the associated cytoplasmic division, cytokinesis.

THE SUBJECT: Late anaphase spermatogonia in the fungus gnat *Sciara*, 30,000X.

BACKGROUND INFORMATION: The German botanist, Hugo von Mohl (1805-1872), gave us the first published description of mitosis in 1837. But, it was not until late in the 19th century when the use of dyes became widespread that the essential facts and terminology of cell division was elucidated. Two patterns of division were detected: mitosis, a conservative process characteristic of somatic cells, and meiosis, a liberalizing process characteristic of division of the germ plasma.

The division of the nucleus, karyokinesis, seems to have dominated the study. It has only been recently that interest in the cytoplasmic aspect of division, cytokinesis, has been stimulated as a possible factor in the very puzzling problem of cellular differentiation. How is it that cells with the same genetic information can differentiate into such divergent tissue forms as the epithelia cell or into the neuron? No generally satisfactory answer has been forthcoming, but the promising lines of investigation equate differentiation to the "turned-on-ness" of only certain genes. Genes might be "turned-on" by their chemical environment which would include first the surrounding cytoplasm of the immediate cell and secondarily the chemical environment of the region in which the cell is located.

The sequence of events observable in the behavior of the nucleus as a cell progresses in division can generally be grouped into four categories, defined generally in respect to the position of the chromosomes. At the onset of nuclear division the chromatin net begins to twist and separate into distinct bodies called chromosomes. The nucleolus wanes, and in higher plants and animals the nuclear membrane resolves itself into a series of vesicles. The bright field microscope shows the chromosome to consist of two major, twisted, thread-like strands. The strands are bound by a single body called the centromere. A series of microtubules and filaments begin to appear, forming a polar axis through the now distinct chromosomes.

In animal cells with centrioles, the centriole pairs drift apart to occupy positions at the poles. The spindle fibers do not originate from the centrioles, but appear to converge at an area outside the centriole body designated as a satellite body. The events to this point characterize the division stage called prophase. When the chromosomes take up an ordered position midway between the spindle poles with the centromeres attached to the fibers of the spindle, the phase is called metaphase. The centromeres then divide, if it is mitosis, creating two new chromosomes. The chromosome body then follows the centromere as it is moved (pushed or pulled) to the pole position. This stage is called anaphase. The chromosomes begin to uncoil, the nuclear membrane reorganizes, the fibers appear, and either a cleavage furrow (animal cells) or a cell plate forms which cleaves the cytoplasm. Two new identical cells are formed. The events of this stage constitute the telophase. The intervals between observable division activity has erroneously been called the "resting state." The better term is interphase or vegetative stage.

Meiosis is associated with the formation of sex cells. Sex cells carry only one of each pair of chromosomes. The events of meiosis are essentially the same as those of mitosis except there is a single duplication of chromosome pairs, and two subsequent cleavages. At the first metaphase in meiosis the analagous chromosome pairs oppose each other on the equatorial plane. The centromeres fail to divide, the consequence of which is the separation of the analagous pairs of chromosomes at the first cleavage. The second cleavage sees the division of the centromeres and the separation of the chromosomes such that each new cell will now contain only half the number of chromosomes characteristic of the species. It is clear that the visible cause of cell division is the replication of the nuclear material. What is not clear is the cause of nuclear material replication. The generally accepted scheme for this is built around the synthesis of the visible double strand of the chromatid and the Watson-Crick model of DNA. Some chemical mechanism or complex of chemical mechanisms trigger the synthesis of an enzyme that causes the double stranded helix to "unzip." This "unzipping" triggers a second enzyme system which utilizes the specificity of the purine-pyrmidine base-pairing and synthesizes a new strand to replace the lost strand of the unzipping operation. The result of this is a duplicate of the pre-existing chromosome with a pair of chromosomes both half old nucleotide and half new nucleotide, held together with a centromere.

The events associated with cytokinesis and the replication of cytoplasmic organelles is still in the preliminary investigative stage. There is speculation that a number of the organelles are self replicating. The general conviction that replication must be associated with the presence of nucleic acids has dimmed promise of self-replication for all the organelles except the mitochondria and chloroplasts, both of which evidence nucleic acids. The self-replicating action of the centrioles is rapidly losing support, mainly because there is no evidence of associated nucleic acid.

The hope for improved understanding of the chemical detail of cell replication will depend on cytochemical sophistication and perhaps improvement in preparations for the electron microscope. Science may one day be in a position to direct rather than describe cell division.

FOR FURTHER STUDY: 1. Discuss the implications of Carl Swanson's statement, "One of the most remarkable phenomena of nature, however, is that in all organisms the process of cell division is essentially the same." 2. Find out about some of the criticisms being made of the Watson-Crick DNA.

Page 7A — Mitosis

Have students complete this page.

A. — prophase (spindle)
B. — telophase
C. — prophase (chromosome-centromere)
D. — interphase
E. — metaphase (spindle fibers)
F. — anaphase
G. — telophase
H. — interphase (nuclear membrane, nucleolus, and centriole)

NOTE — Terms in parentheses refer to the parts indicated by the arrows.

*Page 8 — CENTRIOLE AND FLAGELLUM

CONCEPT: The overwhelming similarity in organization leaves little doubt that the two most fibrillar cytoplasmic organelles, the centriole and the cilium-flagellum, originate from the same primitive organelle, the kinetosome, or basal body.

THE SUBJECT: Flagellum of Chinese hamster sperm, 105,000X. Centriole of weevil primary spermatocyte 85,000X.

BACKGROUND INFORMATION: Cilia and flagella were observed by microscopists studying the protozoa early in the 18th century. It was not until improved microscopes and staining became widespread that the microscopists were able to resolve an axial filament and to note the presence of several filamentous sub-units.

The concept of the cilium (generally short and numerous) and flagellum (generally long and few) remained static until the electron microscope caught up with them. The exciting discovery that there was no fundamental structural difference in cilia and flagella, wherever they appeared, provided an additional unifying theme for living things. The theme for variation of cilia and flagella lies in the basal body and the presence of additional dense fibers in mammalian sperm flagella. The structure of the cilium-flagellum process consists of an outer circle with nine double tubules, and a single pair of central tubules. This is variously referred to as the 9-2 or the 11 fibril structure. The two central filaments are circular in appearance and are not fused. The beat motion of the motile process is at a right angle to the line connecting the centers of the central pair of tubules. The outer doublets are fused. The denser of the sub-units is designated as tubule A. The A tubule has two short arms which point toward the B sub-unit of the next doublet. The significance of this universal theme is not known; nor is the method whereby motion is accomplished known.

Centrioles were discovered by bright field microscopists in association with their study of cell division. During the cell division interphase they become quite distinct. Depending on the plane of section, they may appear as a spherical structure or as two parallel short rods. It was the two-rod appearance that resulted in the term diplosome being applied to the centriole. A line drawn connecting the centrosome and the cell nucleus defines the cell axis. Centrioles are quite frequently found in association with the Golgi complex. They

have been considered to be self-replicating for years. However, careful chemical analysis has failed to turn up any nucleic acid and has produced a large group of skeptics. The tendancy is to hold to the dictum that nucleic acid is an essential prelude to self-replication. Centrioles most frequently occur in pairs. Generally the main axis of one centriole will lie in a plane perpendicular to the main axis of the other centriole. The study of the centriole by the electron microscope has revealed an amazing similarity in fundamental structure to that of the cillium-flagellum. The circle of fibrils contains 9 units, but there are three sub-units per cluster as opposed to two in the cilium-flagellum unit. There is no central pair of tubules in the centriole. The sub-units are named from inside to outside, sub-unit A, sub-unit B and sub-unit C respectively. Sub-unit A again has two short arms. One arm is radially oriented, and the second arm connects to the C sub-unit of the next fibril cluster. It is uncommon to find a cross-section in which the fibrils and the sub-units are all equally distinct. This has led to speculation that the fibrils are not straight but are pitched in a helix. Both the centriole and the motile cell processes originate from kinotosomes. The flagellum of the mammalian sperm clearly originates from a centriole, reflecting a reversibility potential inherent in centriole. This unity seems to be an inherent fact in every natural scheme.

FOR FURTHER STUDY: 1. Find out how the cilia may be varied in presentation in the ciliate protozoan. 2. Find out about some of the suggestions offered for the motion of a flagellum. 3. Find out about the relationship between the centriole and the satellite bodies in which the mitotic spindle fibers terminate.

Page 8A — Centriole and Flagellum

Students are to complete this page, using transparency 8 as a guide.

*Page 9 — MITOCHONDRIA

CONCEPT: The mitochondria is the energy modulator, the modulator, the powerhouse of the cell.

THE SUBJECT: Salivary gland of the fungus gnat, *Sciara*, 70,000X.

BACKGROUND INFORMATION: Rod-like structures, known now to have been mitochondria, were reported by brightfield microscopists during the later 19th century following the widespread use of stains and the perfection of the substage condenser and immersion lense. They were first called bioclasts and thought to be bacteria-like.

The name mitochondria (thread-like form) was first used by Benda in 1898. An opportunity for the elucidation of some of the function of the mitochondria was provided by Otto Warburg in 1920 with his discovery that most of the cytoplasmic oxidative reactions were confined to the large granular fraction of his centrifuged homogenate. It was not until the middle 1930's that Warburg's discovery was put to use and the biochemistry of the mitochondria started to unfold. Using a refined Warburg technique, the mitochondria were demonstrated to be the site of active oxidative enzymes

BIOLOGY — Cell Machinery

and the sole cellular depository of some enzymes. It is generally conceded that the Kreb's cycle is a mitochondrial phenomenon. As such, the lions share of ATP is secreted by the mitochondria. Glycolysis proceeds in the cypoplasm to the oxidation of 3-phosphoglyceraldehyde. All reactions beyond this occur in the mitochondria.

The morphological character of the mitochondria had to await the refinement of fixing and thin section techniques to a level satisfactory for electronmicroscopy. Palade and Sjostrand in 1953 independently discovered that the mitochondria had an elaborate membrane structure. The most consistent feature of the sausage-shaped organelle is the double membrane. The outer membrane usually has a smooth enveloping contour. The inner membrane has an exaggerated surface being thrown into shelf-like, or tube-like folds which extend into the interior. The tube-like folds are more typical of plant mitochondria, while the shelf-like folds are more representative of animal mitochondria. These inner folds are called cristae mitochondriales or just cristae. It has been generally believed that the enzymes of the Kreb Cycle are located in the region between the membranes and the electron transport mechanism in the membrane proper. But some later electron micrographs, using a "negative staining" technique, revealed stalked particles with a hollow stem and spherical head spaced at regular intervals along the inner membrane. These structures may bear the oxidative phosphorylation enzymes.

Mitochondria frequently show a pattern in their address to the endoplasmic reticulum. An arm of the endoplasmic reticulum will approach a mitochondrial body, fall into a locus parallelling the mitochondria membrane, and proceed for a considerable length of the body. The significance of this, if it is indeed significant, is not known.

The mitochondria is reported to contain small amounts of nucleic acid. This has made the origin of mitochondria a matter of interest. They may indeed be self-replicating. Mitochondria do experience turnover. There is evidence of mitochondria in various phases of lysis within the lysosomes. Assumedly, this is the mechanism for their disassembly.

Mitochondria have been reported as constant components of all cells except the cells of the Monerans, bacteria and blue-green algae, and the mature mammalian red blood cells. As the principal site of chemical bond energy transformation, the mitochondria may be properly called the ATP generator or the powerhouse of the cell.

FOR FURTHER STUDY: 1. Find out how the level of metabolic activity of a tissue is related to the level of cristae expression in the associated mitochondria present or in the number of mitochondria present in the cells of the tissue. 2. Find out about the speculation that the mitochondria are "little" organisms living symbiotically in the cells of the protozoan, metaphyta, and metazoa.

Page 9A — Mitochondria

Have students label this page according to transparency 9.

*Page 10 — GOLGI COMPLEX

CONCEPT: The Golgi complex is a vacuolar, canalicular organelle associated with secretion and absorption, and with cell division.

THE SUBJECT: Spermatid of Chinese hamster with Golgi complex manifesting its role in spermacrosome formation, 27,000X.

BACKGROUND INFORMATION: The Golgi complex holds the distinction of being the most disputed cellular organelle and also of being the only organelle bearing the name of a cytologist. Camillio Golgi, an Italian cytologist, first reported the structure in 1898 as an internal reticular apparatus observable in elaborately stained nerve cells. The inability to substantiate its existence in living tissues and the elaborate staining used to produce it caused many to accredit it to preparation artifact. However, the use of stains containing osmium or silver established the universal existence of the organelle in fixed material. The phase contrast microscope affirmed its existence in living material in the 1940's. The fine details of its organization were elaborated by the electron microscope in 1952.

The Golgi complex is also called the dictyosome. Its general transverse appearance is that of a series of double membraned discs with small vesicles around the edge. The system of membranes is continuous with the outer nuclear membrane and the smooth endoplasmic reticulum. The series of 4-8 flat membrane saccules are stacked one on another, the whole unit being most frequently curved. Vesicles always hover around the convex outer edge of the cisternae. Using homogenation and differential centrifugation, the Golgi complex has been concentrated in sufficient quantity to provide some insight into its chemical role in the cell. There are some connections of the golgi complex and the membranes of the endoplasmic reticulum. The rough type of endoplasmic reticulum secretes certain substances which are passed to the golgi bodies. Here they are processed and then packaged into the vesicles. The vesicles move to the plasma membrane and release their contents outside the cell. The golgi bodies also produce lysosomes, the small membranous bags which contain enzymes.

The origin of the organelle is not clear. The absence of nucleic acid and its phospholipid membrane composition places doubt on any self-replicating role. A mechanical scheme for its elaboration from the endoplasmic reticulum is intuitively attractive, but stands without adequate supporting evidence. Thus the Golgi complex is solidly established in the cell's repertoire as the agent of import and export. It may play other roles as well.

BIOLOGY — Cell Machinery

FOR FURTHER STUDY: 1. Find out why there was so much concern regarding artifact in the early work with the electron microscope. 2. Find out about some of the suggestions that have been made regarding the mechanics whereby the Golgi complex aids in importing and exporting material through the plasma membrane.

Page 10A — Golgi Complex

Students are to complete this page according to transparency 10.

*Page 11 — ENDOPLASMIC RETICULUM
WITH RIBOSOMES

CONCEPT: Ribosomes are the ubiquitous cytoplasmic organelles charged with the responsibility of protein synthesis. The endoplasmic reticulum has been called the cytoskeleton.

THE SUBJECT: Pancreas cell of the Chinese hamster, 25,000X.

BACKGROUND INFORMATION: Brightfield microscopists got hints of the existence of a cytoskeleton network and certainly resolved a granular and a smooth kind of cytoplasm. But the definition was not adequate to reveal the true nature of the two organelles they were seeing.

The discovery of the endoplasmic reticulum (ER) was truly one of the exciting moments in electron microscopy. The interior of the cell was not a monotonous sol, but a complex of anastomosing membranes permeating the cytoplasm and enveloping the nucleus. The ER is also known as the ergastoplasm. It consists of an elaborate series of narrow canals delineated by a single membrane. The single membrane is continuous with the outer nuclear membrane.

The resolution that brought out the ER also revealed a cytoplasmic granule, the microsome, frequently arranged along the outer sides of the ER channels. A chemical analysis of these granules revealed that they had a high RNA content. The name was revised consistent with this to ribosomes. The ribosome is known to be a constant component of every cellular organism including the monerans. When ribosomes exist in association with the ER, it is called rough ER. The ER without ribosomes is called smooth ER. The former situation prevails in cells engaged in the synthesis of protein, the latter in cells engaged in the synthesis of fatty substance.

High resolution of the ribosome reveals its hemispherical structure. Chemical analysis has revealed that the ribosome consists of two sub-units held together by electrostatic forces. The electrostatic forces can be relieved and the ribosome broken into an RNA sub-unit and a protein sub-unit. The RNA sub-unit accounts for 40-60% of the ribosome. There are three known categories of RNA. The RNA residing on the ribosome is called ribosomal RNA and appears to be very constant. There is a messenger RNA emanating from the nucleus which associates with the ribosomal RNA in some

manner facilitating protein synthesis. A third kind of RNA is called transfer FNA. Emanating from the nucleus, it bonds with specific amino acids in the cytoplasm needed by the ribosome in protein synthesis.

When the amino acid is delivered to the ribosome, it is incorporated into the protein being synthesized. Conceptually the ribosome is said to "read" the messenger RNA and in turn understand what protein the nucleus wants made. It then makes the protein using the pieces supplied by the transfer RNA.

This is certainly dramatically descriptive, but a chemical rationale for this is something else. It is a fact the ribosome is the cytoplasmic site of protein synthesis. There is considerable evidence than ribosomes are generalists, being able to synthesize most any protein, so long as it has messenger RNA instructions, and the essential amino acid pieces. Based on the Watson-Crick model of DNA, it is assumed that the nuclear RNA consists of copies of DNA using the uracil substitution for thiamine and the ribose sugar for deoxyribose sugar. How instructions are given regarding what part of the DNA to copy is not known.

The ER is not essential for the effective protein synthesis of the ribosomes. The ER is lacking in the bacteria, yet the bacterial ribosomes are as efficient as any other ribosomes in protein synthesis. Where the ribosomes are related to the ER a likeness has been drawn to navigable harbors (the ER channels) and the wharf (the ribosomes). Cooperatively, the ER and the ribosome contribute heavily to the action of the cell. The ER serves as a cytoskeleton, a surface for chemical reactions, and a channel for transfer and communication. The ribosomes provide the knowhow in making peptid bonds. They are most abundant in the cytoplasm but also appear in the nucleus.

FOR FURTHER STUDY: 1. Find out how the invasion of a virus affects the protein synthesis of a cell. 2. Find out what the cellular source of the non-essential amino acids is.

Page 11A — Endoplasmic Reticulum and Ribosomes

Students are to complete this page, using transparency 11 as a guide.

*Page 12 — PLASMA MEMBRANE

CONCEPT: All the membranes of the cell have the same trilaminar structure, but the unity of structure belies the variability of property.

THE SUBJECT: Salivary gland of the fungus gnat, 95,000X.

BACKGROUND INFORMATION: The brightfield microscopists were able to observe the coarse outline of the plasma membrane by subjecting cells to hypertonic solutions. But their resolving power was insufficient to provide any hint of the structural detail. Most of the behavioral properties of membranes were developed before the electron microscope showed the unit membrane to consist of three distinct zones. Fat solvents were known to move freely through the plasma membrane. Its lipid content was then known. The presence of a layer of protein to provide the elasticity and explain the

BIOLOGY — Cell Machinery   9

surface tension phenomenon had also been hypothesized. The electron microscope revealed two dense zones separated by a clear zone. Reconciling this visual pattern with the properties known stimulated the current model of the unit membrane as a lipid layer sandwiched between two protein layers. The ubiquity of this pattern wherever membranes were encountered caused it to be termed the unit membrane. Later findings suggest that the lipid filling is really in the form of a double layer.

The plasma membrane functions to maintain the integrity of the cell through the regulation of egress and ingress. The membrane is considered to be a part of the living cell and is therefore capable of changing its character from moment to moment, consistent with the needs of the cell and the demands of the environment. Some of the movement of material through the plasma membrane can be explained in terms of free energy differences, but the greater part of this action is accredited to active transport.

Active transport is an energy consuming action. The mechanism for active transport is poorly understood for most materials, but rather well established for bulk transport. In pinocytosis the membrane invaginates, fills, and then separates from the outside contour forming a vesicle. In phagcytosis the membrane actively engulfs the particle, then separates from the outer contour and becomes disorganized in the face of enzymes which attack the particle.

The mechanisms for active transport of ions and other desired molecules must be worked out in molecular terms. Microscopists can probably be of no help with this problem.

The plasma membrane of adjacent cells are not face to face but are separated by a matrix. The matrix is usually pectin for plants and hyaluronic acid for animals. Very frequently the matrix is bridged by dense bands that appear to suture the plasma membranes together. These fiberous bands are called desmosomes. The unit membrane appears elsewhere in the cell. Vacuoles, lyosomes, endoplasmic reticulum, and Golgi complex are all delineated by a single unit membrane. The chloroplast, the mitochondria, and the nucleus are all bound by a double membrane, that is, two juxtaposed unit membranes. The membrane then acts as a skin for the cell. It is variable in properties, and it serves to regulate what enters the cell and what escapes from the cell.

FOR FURTHER STUDY: 1. Find out about the theories offered for active ion transport across the membrane. 2. Find out about the theories offered for the active transport of sugar across the membrane.

Page 13 – UNIT TEST ON THE CELL

1. D   2. C   3. B   4. C   5. B   6. D   7. A   8. A
9. B   10. C.

Page 14 – (Unit Test continued)

11. A   12. B   13. A   14. B   15. C   16. C   17. C
18. C   19. B   20. A   21. A   22. B.

Page 15 – (Unit Test continued)

23. D   24. B   25. B   26. C   27. A   28. C   29. C
30. C   31. B   32. C   33. A.

Page 16 – (Unit Test continued)

34. A   35. A   36. C   37. C   38. C   39. B   40. A
41. A   42. C   43. A   44. C   45. C.

BASIC MICROSCOPES FOR CELL STUDY

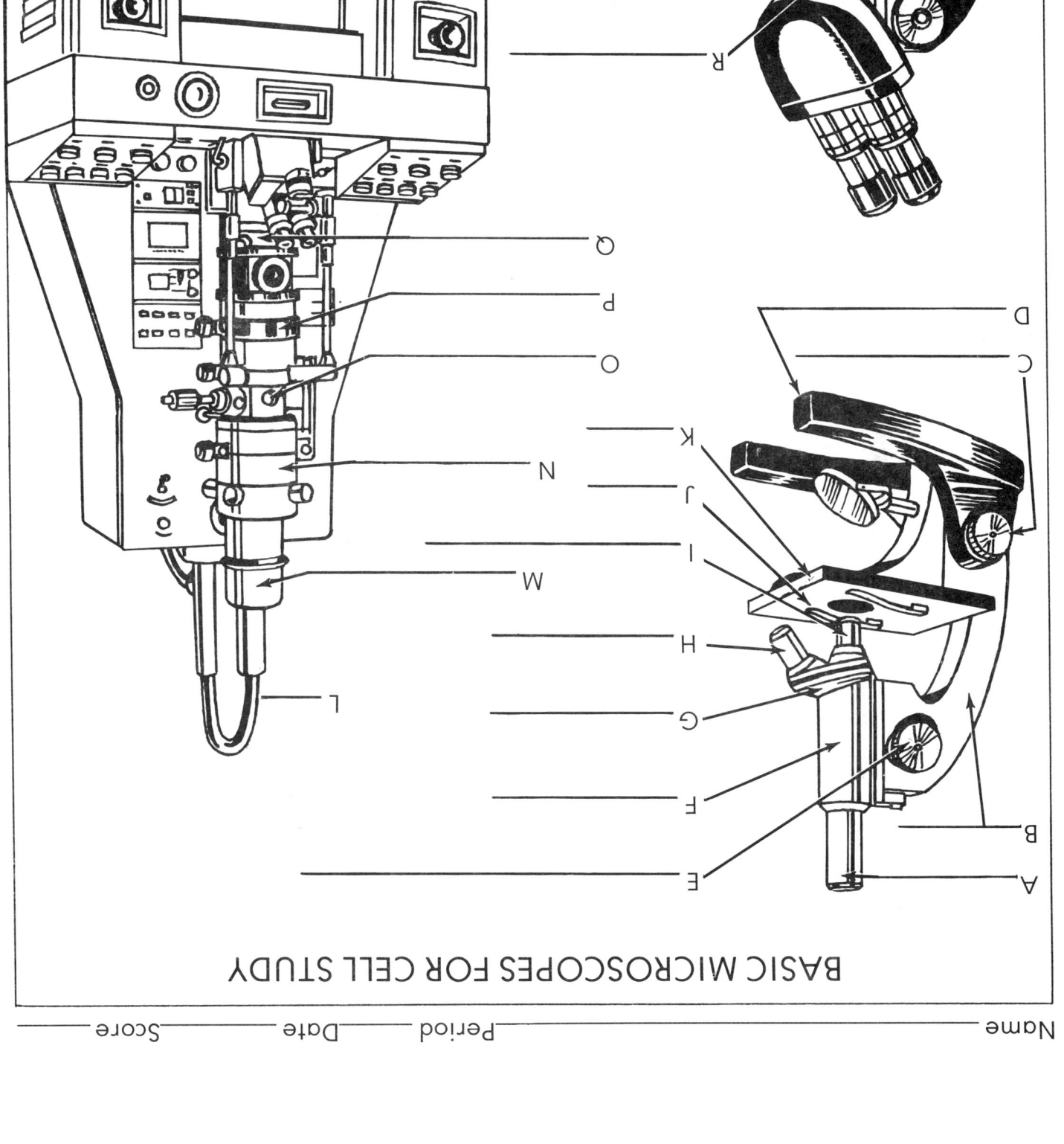

CELL MACHINERY

A PLANT CELL

3A
MILLIKEN PUBLISHING CO.

Name _____ Period _____ Date _____ Score _____

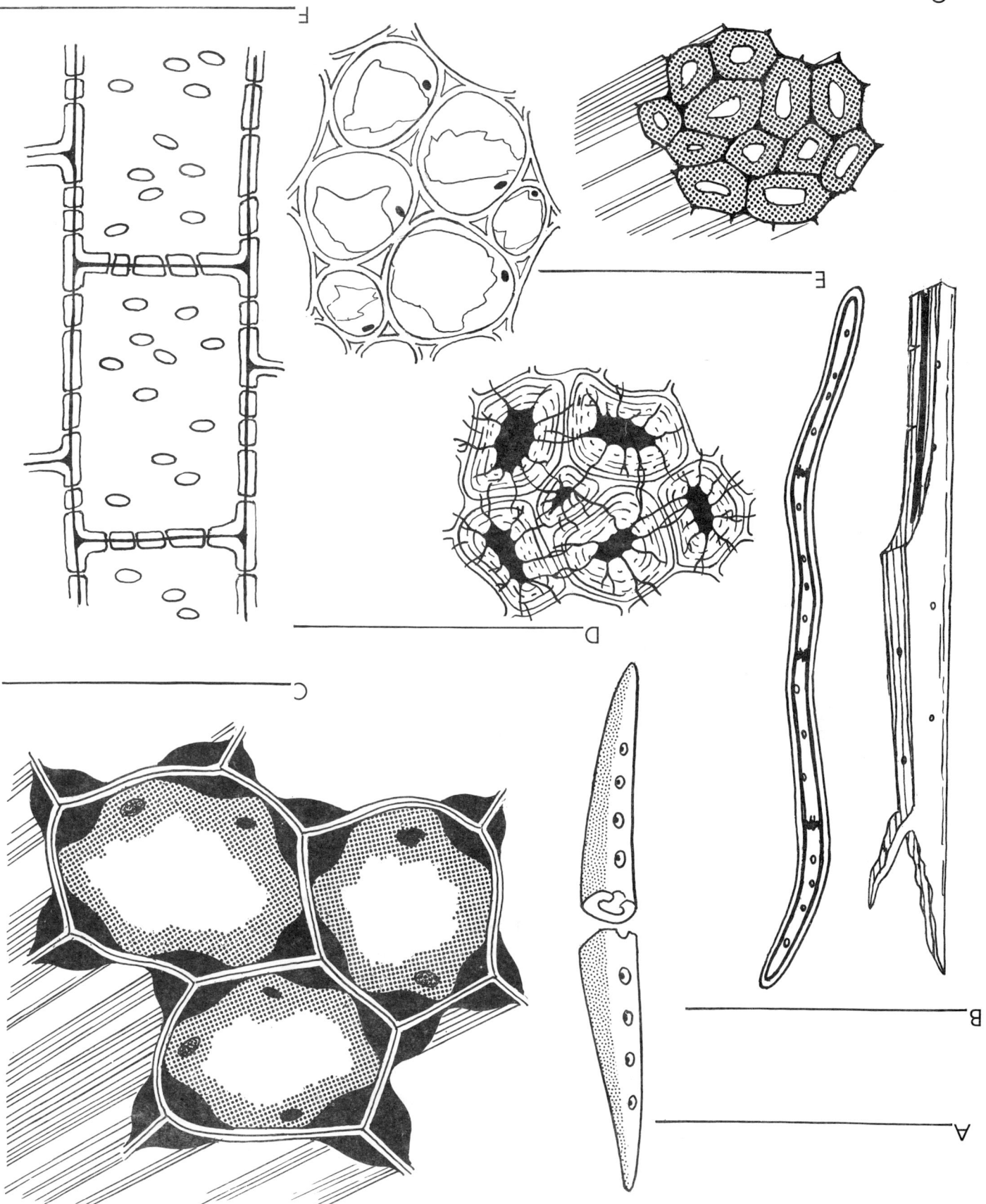

CELL MACHINERY

AN ANIMAL CELL

Name _____ Period _____ Date _____ Score _____ 4A

MILLIKEN PUBLISHING CO.

A _____
B _____
C _____
D _____
E _____
F _____
G _____
H _____
I _____
J _____

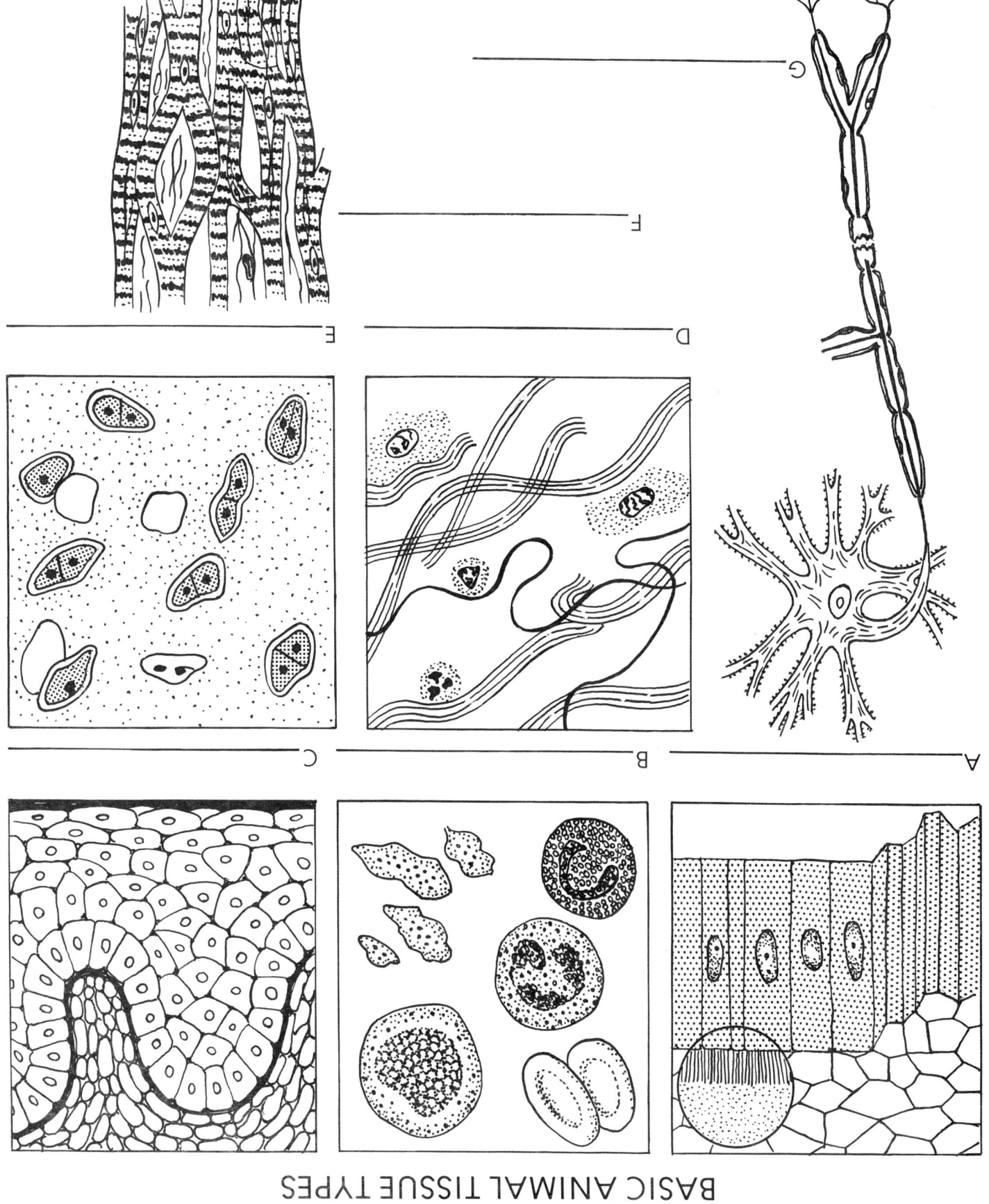

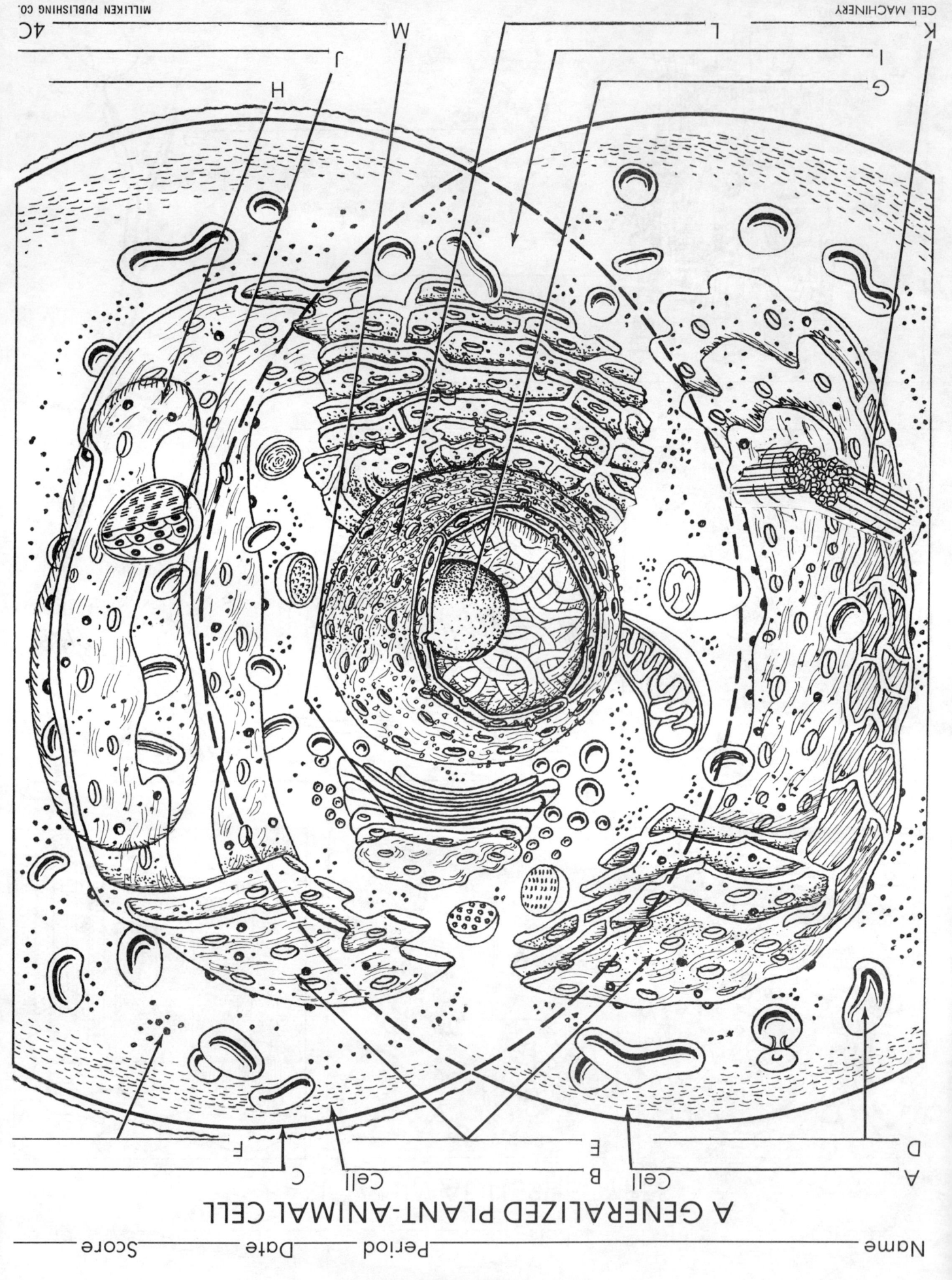

CHROMOSOME REPLICATION

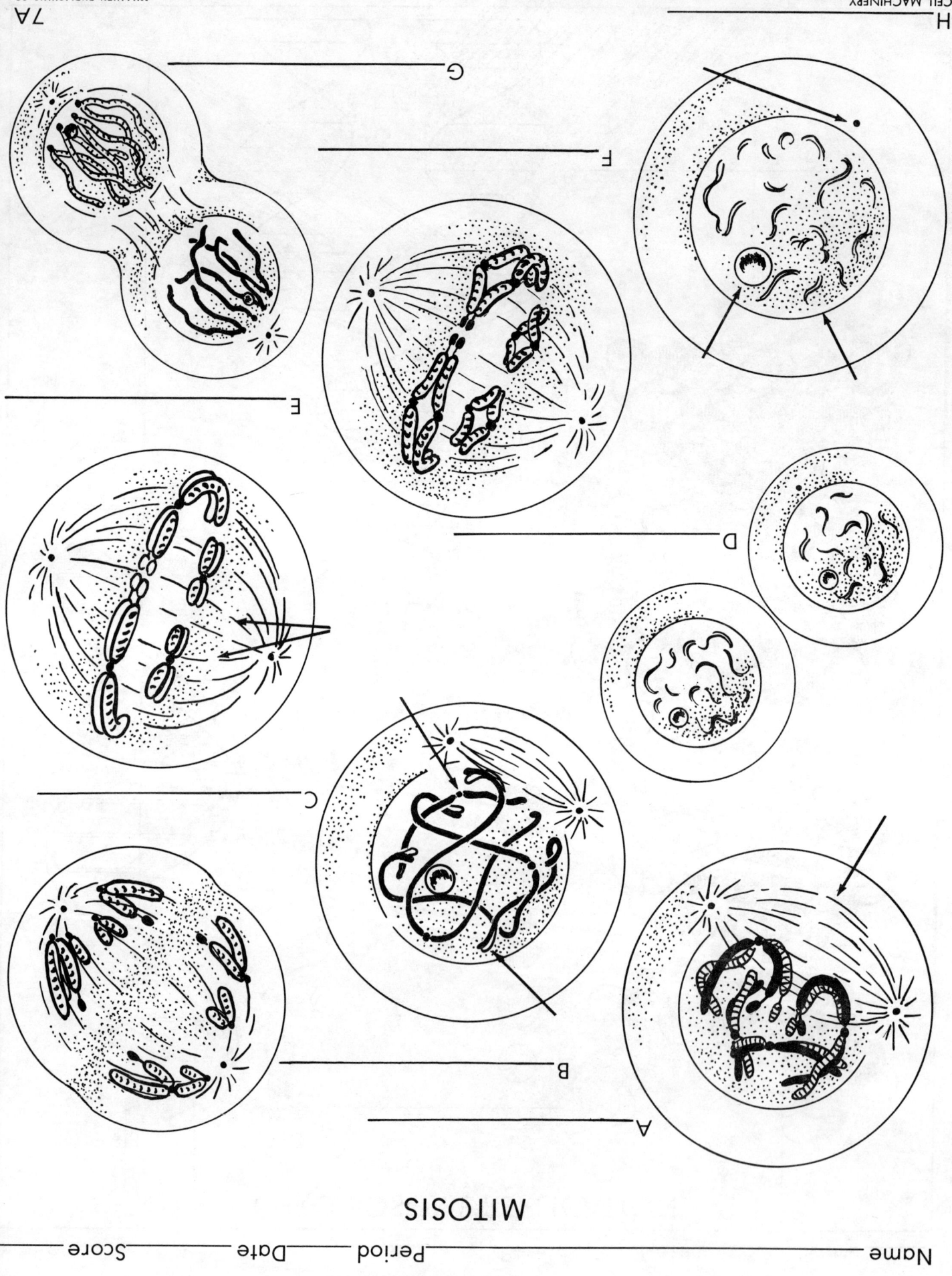

MITOSIS

CENTRIOLE AND FLAGELLUM

Name _____ Period _____ Date _____ Score _____

Name_____ Period_____ Date_____ Score_____

MITOCHONDRIA

A

B

C

D

CELL MACHINERY

9A

MILLIKEN PUBLISHING CO.

CELL MACHINERY

GOLGI COMPLEX

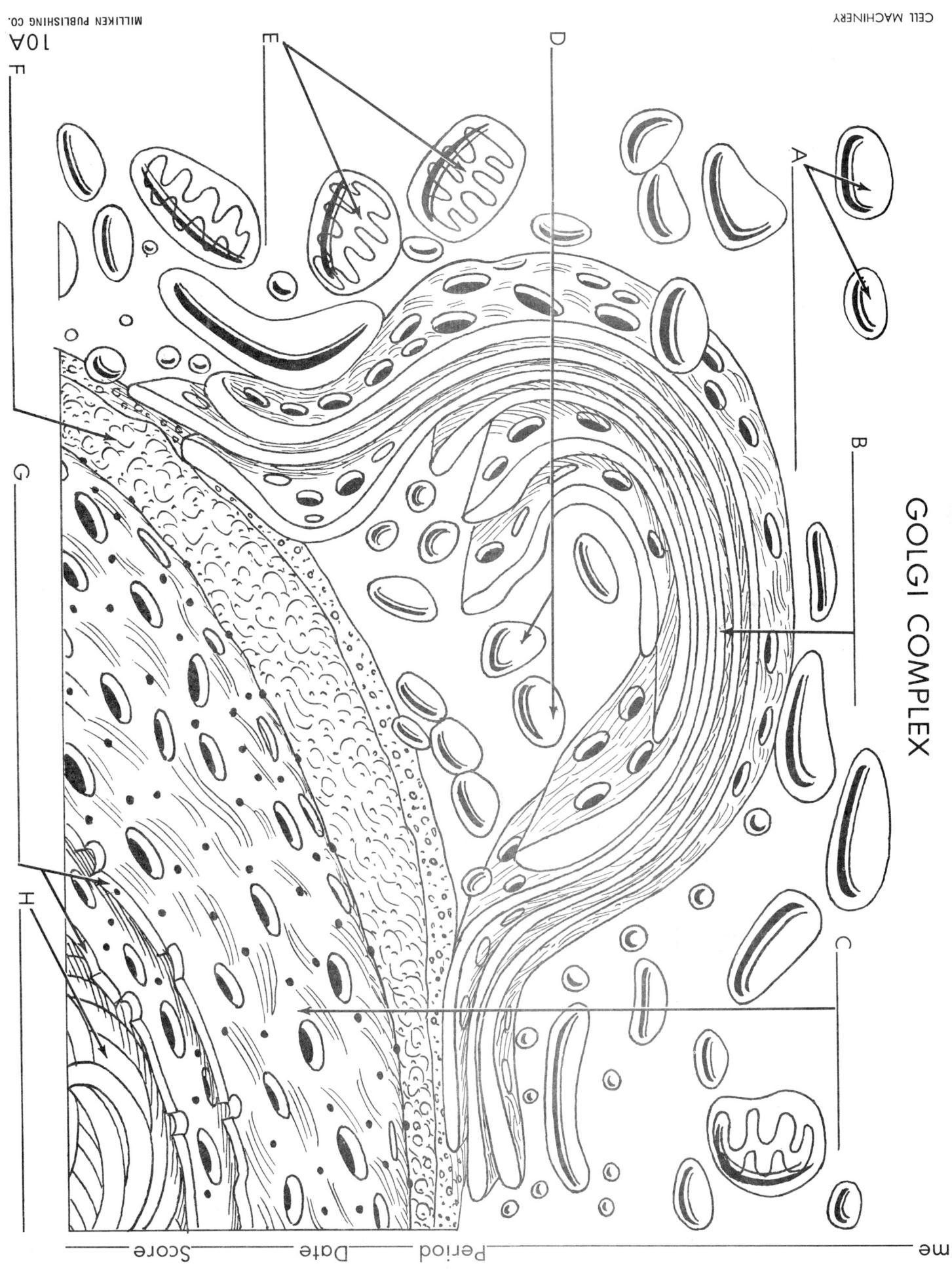

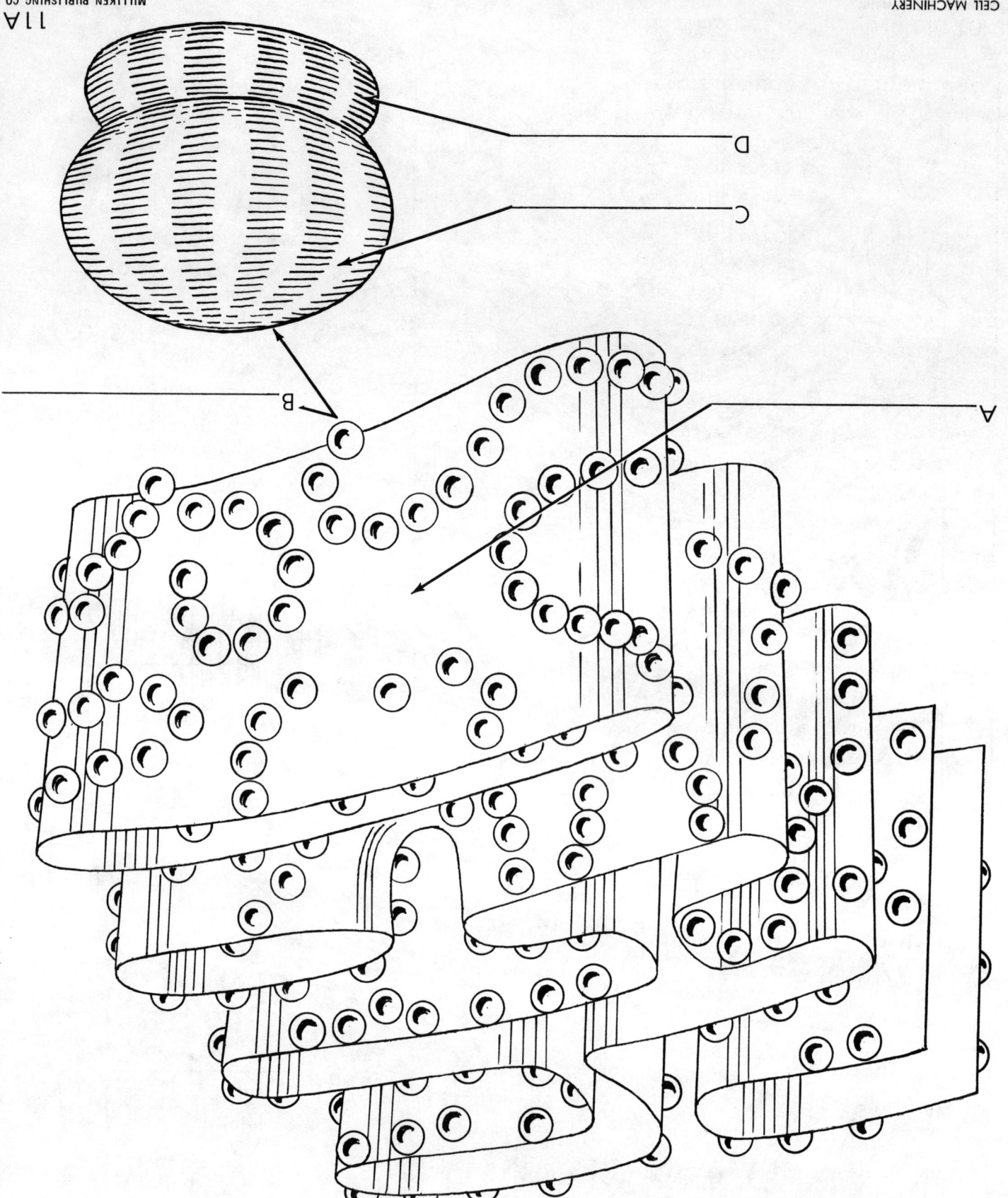

ENDOPLASMIC RETICULUM AND RIBOSOMES

Name _____Period _____Date _____Score _____

UNIT TEST ON THE CELL

Select only one answer for each of the following situations.
Indicate your answer by circling the appropriate letter.

A B C D 1. The most descriptive name
for the unit of living things
would be
A. the cell.   C. the nucleus.
B. the atom.   D. the corpuscle.

A B C D 2. The development of the cell
theory had to await the devel-
opment of the
A. printing press.
B. photographic process.
C. microscope.
D. microtome.

A B C D 3. The maximum resolving
power of the human eye is of
objects with mean diameters of
A. 1000 microns. C. 10 microns.
B. 100 microns. D. 100
angstroms.

A B C D 4. The major contribution of
chemistry to cell study was the
A. phase contrast microscope.
B. electron microscope.
C. coal tar dyes.
D. formaldehyde.

A B C D 5. The principle contribution of
the phase contrast microscope
to cell study was
A. increased magnification.
B. ability to study living
material.
C. definition at the molecular
level.
D. the shifting of light waves a
quarter of a wave out of
phase.

A B C D 6. Focusing the electron microscope is accomplished by
A. evacuating the tube.
B. intensification of the electron beam.
C. turning the coarse adjust-
ment knob.
D. varying the magnetic field
in the objective lens.

A B C D 7. The viewing aspect of the
electron microscope specimen
is
A. a specimen shadow.
B. a specimen reflection.
C. a light image of the
specimen.
D. an inverted image of the
specimen.

A B C D 8. The biochemical knowledge
of the cellular organelles came
about through the use of
A. differential centrifuging.
B. the electron microscope.
C. phase contrast microscope.
D. use of coal tar dyes.

A B C D 9. Plant cells, in general, are
best characterized by the
presence of
A. chlorophyll.
B. cellulose cell wall.
C. centrioles.
D. lysosomes.

A B C D 10. Which of the following
terms includes all of the others?
A. ribosome   C. nucleus
B. chromatin   D. nucleolus

13

CELL MACHINERY

MILLIKEN PUBLISHING CO.

Name _____ Period _____ Date _____ Score _____

A B C D 11. A nucleus imbedded in cytoplasm best reflects Andre Lwoff's unity of
A. plan.   C. composition.
B. function.   D. parts.

A B C D 12. Usually the cell's largest organelle is the
A. mitochondria. C. plastid.
B. nucleus.   D. endoplasmic reticulum.

A B C D 13. A membranously bound nucleus is absent from the representatives of the Kingdom
A. Monera.   C. Metaphyta.
B. Protista.   D. Metazoa.

A B C D 14. Judging by the kind of stain to which it responds most readily, the nucleus is evidently
A. basic.   C. fatty.
B. acidic.   D. starchy.

A B C D 15. The nucleolus is morphologically distinguished by the
A. double membrane.
B. abundance of ribosomes.
C. distinct fibers.
D. dense staining.

A B C D 16. The protein associated with the nucleus is
A. DNA.   C. histone.
B. RNA.   D. nucleic acid.

A B C D 17. The most exact term for describing the replication of the nucleus material is
A. cell division.
B. cell replication.
C. karyokinesis.
D. cytokinesis.

A B C D 18. The locus concept of genes on chromosomes was given visual support by the
A. nucleotide pairing.
B. the double helix.
C. dark and light bands in some chromosomes.
D. tangled nature of the chromatids.

A B C D 19. Chromosome maps were developed using information from
A. nucleotide pairing.
B. crossover frequencies.
C. the phase contrast microscope.
D. the electron microscope.

A B C D 20. The double helix concept of DNA was supported by the 1:1 nucleotide pairing between thymine and
A. adenine.   C. cytosine.
B. uracil.   D. guanine.

A B C D 21. Which of the following is not a way DNA differs from RNA?
A. Its sugars are bonded by phosphate groups.
B. The sugar used is missing an oxygen.
C. Uracil is noticeably absent
D. It is confined to the nucleus.

A B C D 22. Mitosis and meiosis are terms used to describe
A. the division of the cell.
B. the division of the nucleus.
C. the division of the cytoplasm.
D. cytokinesis.

CELL MACHINERY

MILLIKEN PUBLISHING CO.

Name _____ Period _____ Date _____ Score _____

A B C D 23. The chromosomal structure that moves the chromosome along the spindle fiber during karyokinesis is called the
A. cilia.　　　C. centriole.
B. flagella.　　D. centromere.

A B C D 24. When the chromosome becomes aligned on the equatorial plate of the mitotic spindle, the phase of division' is called
A. prophase.　　C. anaphase.
B. metaphase.　D. telophase.

A B C D 25. The appearance of a particular chromosome is caused by its own size and
A. the double helix.
B. the location of the centromere.
C. the mitotic spindle.
D. the asterol apparatus.

A B C D 26. The separation of the chromosomes and advance away from the equatorial plate marks the inception of the
A. prophase.　　C. anaphase.
B. metaphase.　D. telophase.

A B C D 27. The cleavage furrow for the animal cells
A. moves in from the outside.
B. spreads from the middle out.
C. quantitatively divides the cytoplasm.
D. causes the nuclear membrane to reform.

A B C D 28. It is generally felt that cytoplasmic organelles capable of self replication must have
A. a double membrane.
B. their own ribosomes.
C. some nucleic acid.
D. a centromere.

A B C D 29. A good microscopist could most likely not tell the difference between a cross section of a flagellary process and a a cross section of a process of
A. the spindle fibers.
B. the nucleolar fibers.
C. the cilia.
D. a centriole.

A B C D 30. The beat motion of a flagella is at a right angle to a line connecting
A. subunit A & B.
B. subunit A & C.
C. the central tubules.
D adjacent flagella.

A B C D 31. A cellular organelle most commonly found in association with the centriole is the
A. endoplasmic reticulum.
B. Golgi complex.
C. ribosome.
D. mitochondria.

A B C D 32. The tubule arrangement seen in a cross section of a centriole is
A. 9-2.　　C. 9 x 3-0.
B. 11-0.　　D. 9 x 3-2.

A B C D 33. Adjacent centrioles are most commonly
A. at right angles to each other.　C. obliquely arranged.
B. parallel to each other.　D. arranged in rows.

CELL MACHINERY

MILLIKEN PUBLISHING CO.

Name _____ Period _____ Date _____ Score _____

A B C D 34. The most distinguishing morphological feature of a mitochondria is its
A. elaborate membrane system.
B. concentrations of ribosomes.
C. enzyme-bearing projections.
D. variable size and shape.

A B C D 35. Which of the following processes is almost entirely a mitochondrial phenomenon?
A. Kreb's cycle
B. glycolytic cycle
C. protein synthesis
D. carbon fixation cycle

A B C D 36. The internal folds of the mitochondria are called
A. septa.
B. ER.
C. cristae.
D. vesicles.

A B C D 37. Mitochondria are seriously suspected of being self-replicating, mainly because of
A. the double membrane.
B. the abundance of ATP.
C. the presence of nucleic acid.
D. their large size and complexity.

A B C D 38. There is evidence that some animal cells can dispose of mitochondria. The organelle doing this lysis work is the
A. ribosome.
B. vacuole.
C. lysosome.
D. plastid.

A B C D 39. The basic chemical usefulness of the mitochondria lies in their secretion of
A. protein.
B. ATP.
C. sugar.
D. lipids.

A B C D 40. Mitochondria are present in the cells of all organisms except the representatives of the Kingdom
A. Monera.
B. Protista.
C. Metaphyta.
D. Metazoa.

A B C D 41. The Golgi complex can best be identified by
A. its stacked membranous discs.
B. the presence of the centriole.
C. its association with the ER.
D. its association with the nucleus.

A B C D 42. The most certain role of the Golgi complex is that of
A. protein synthesis.
B. fat synthesis.
C. secretion and absorption.
D. cellular metabolism.

A B C D 43. The endoplasmic reticulum might best be described as the
A. cytoskeleton.
B. synthesis center.
C. power house.
D. nerve center.

A B C D 44. The most puzzling type of RNA is the
A. messenger RNA.
B. transfer RNA.
C. ribosomal RNA.
D. nuclear RNA.

A B C D 45. A cell could best be defined as
A. a unit of organization.
B. a unit of structure.
C. a work area for a complex of organelles and macromolecules.
D. a living thing.

CELL MACHINERY

MILLIKEN PUBLISHING CO.

16

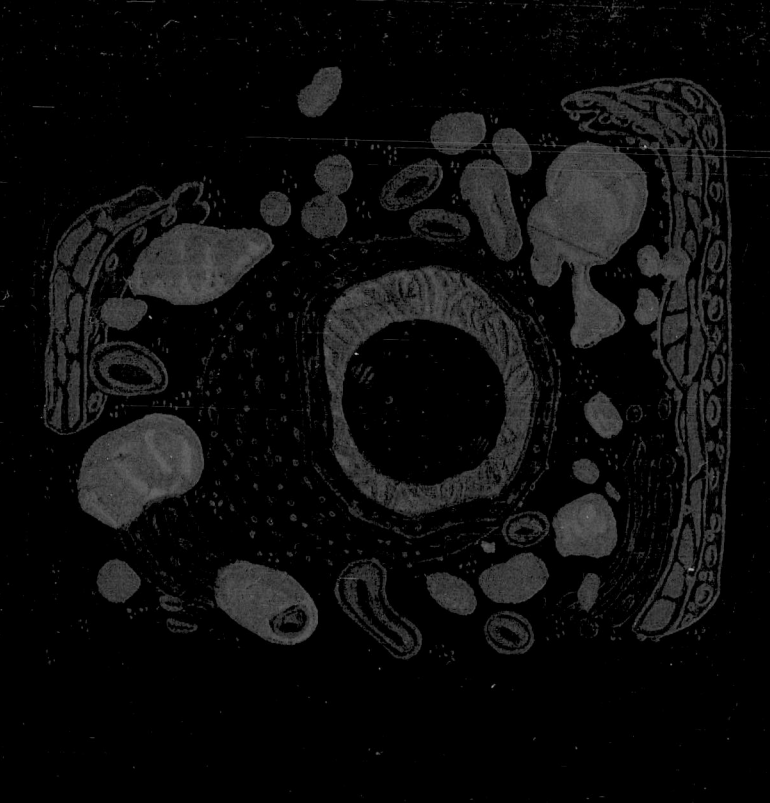

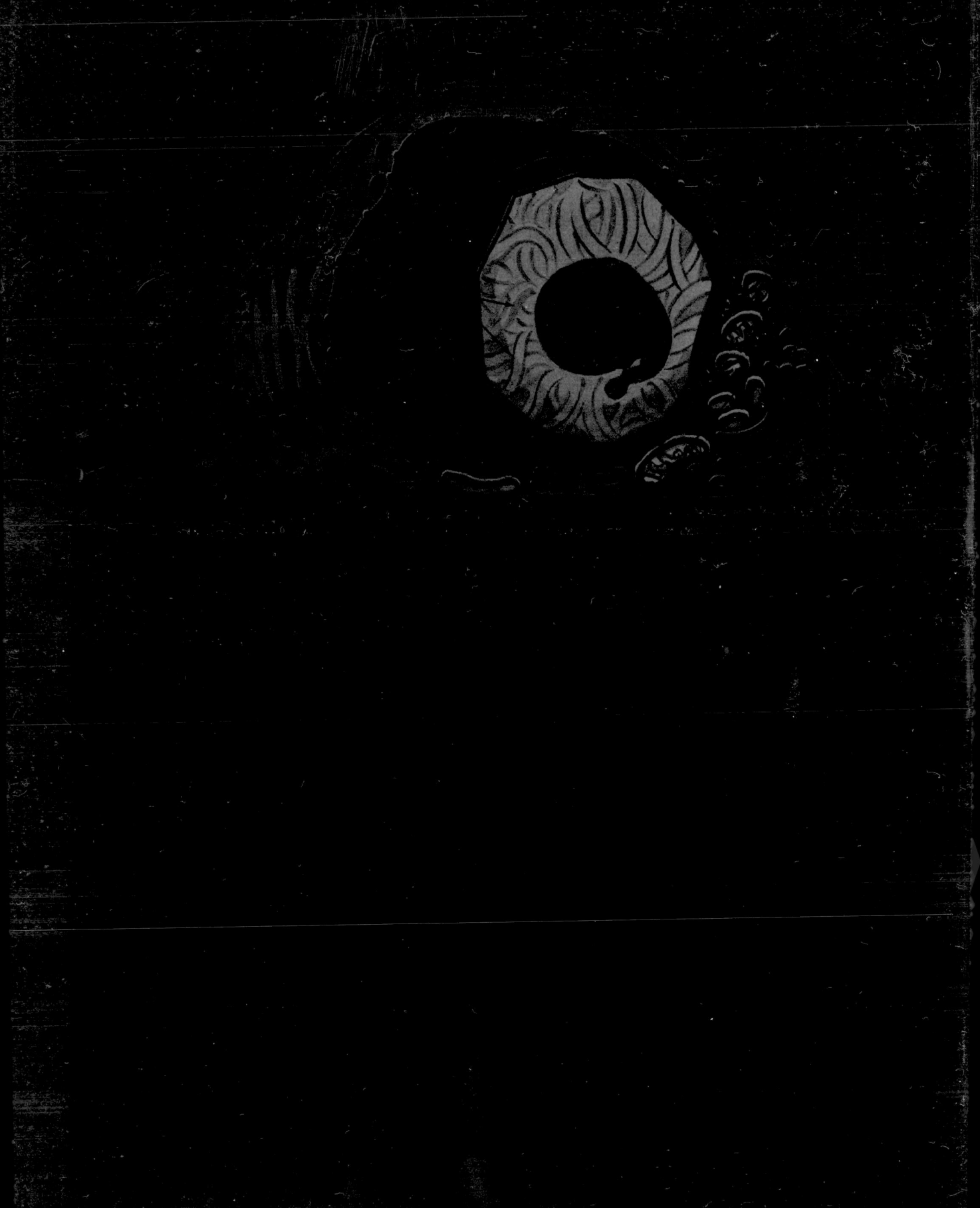

MP-4721 - 500
MP-4724 - 1050 - old
MP-4725 - 300
MP-4730 - 700
MP-4733 - 1100
MP-4753 - 800
MP-4762 - 1000
MP-4763 - 1700
MP-4767

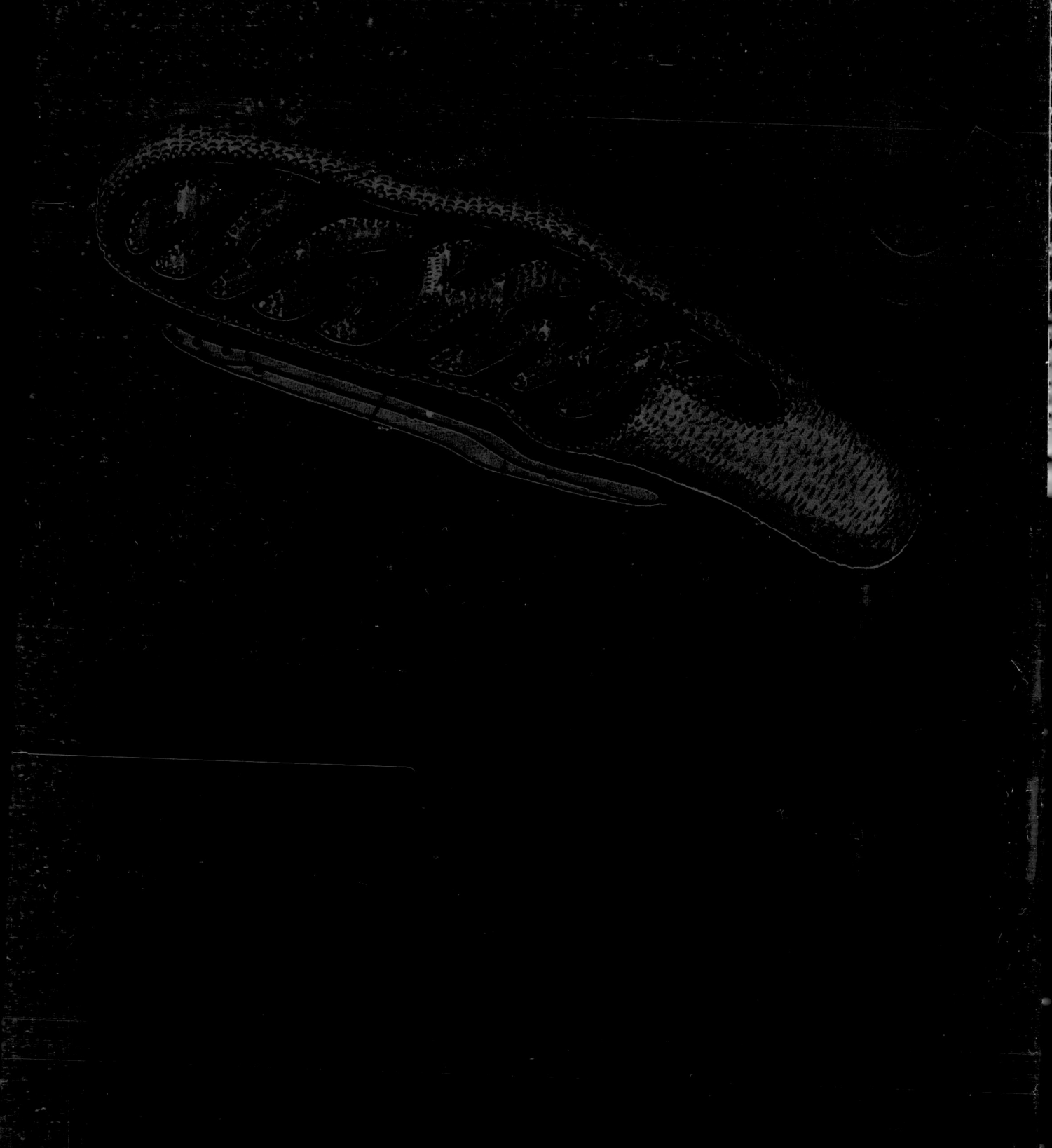

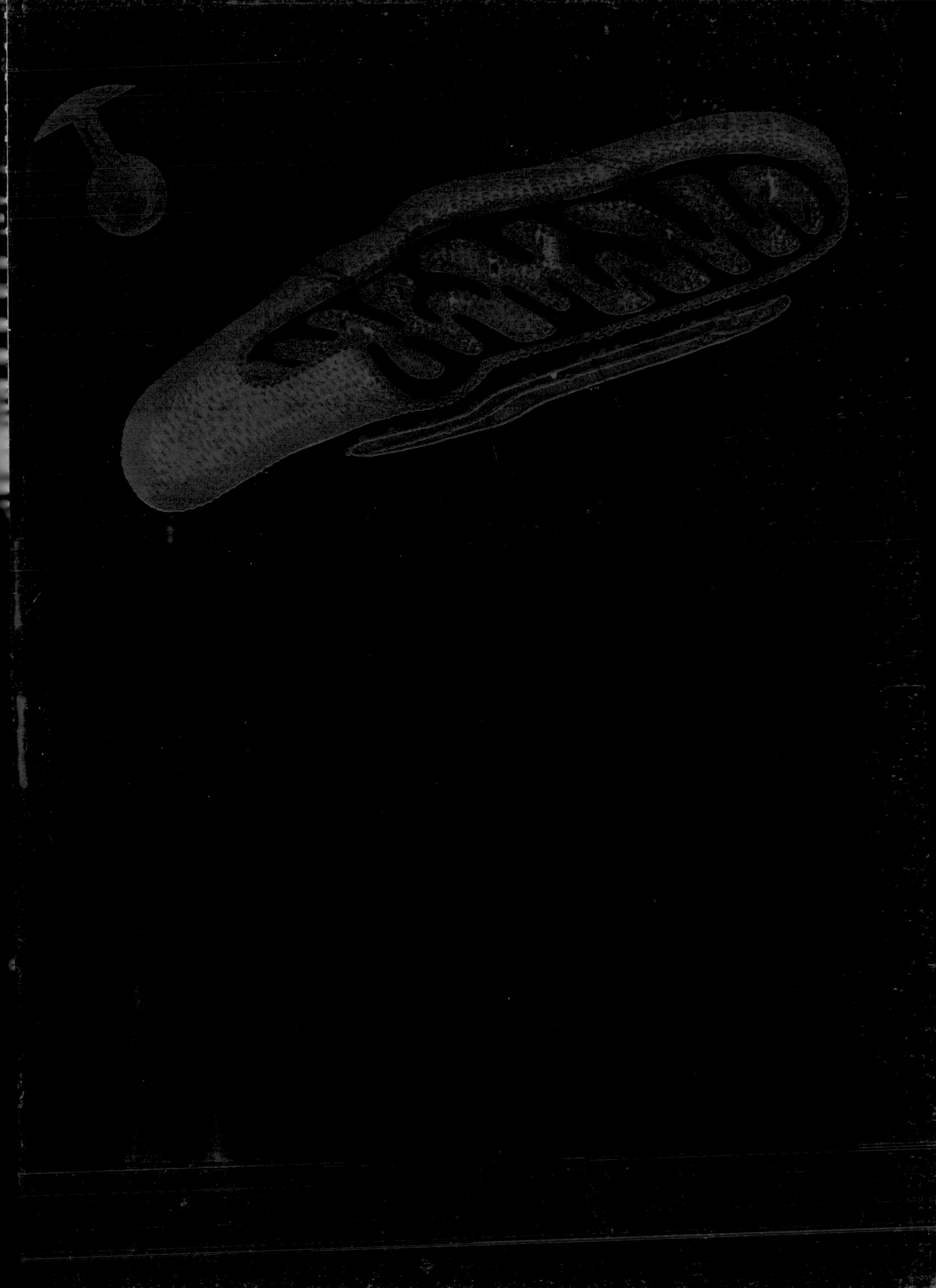

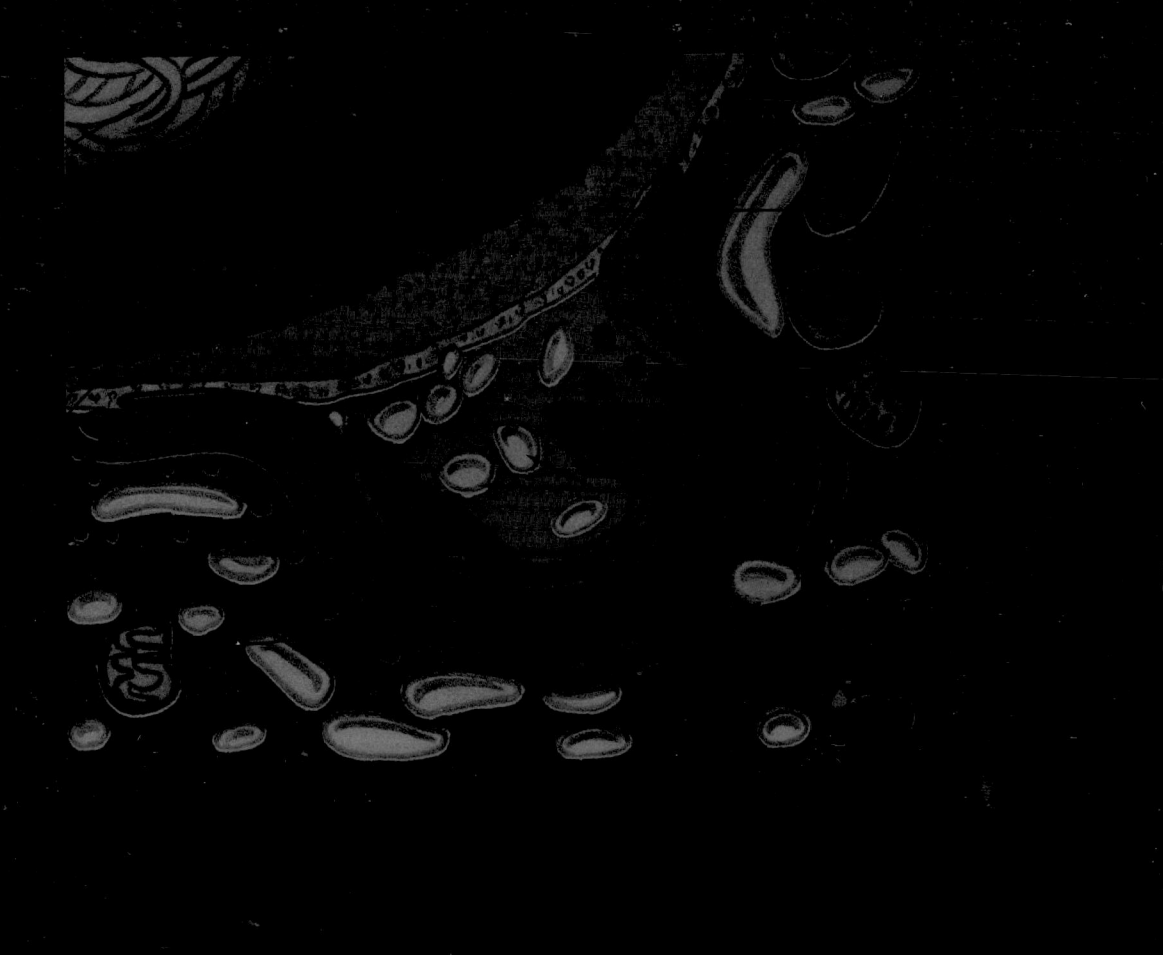

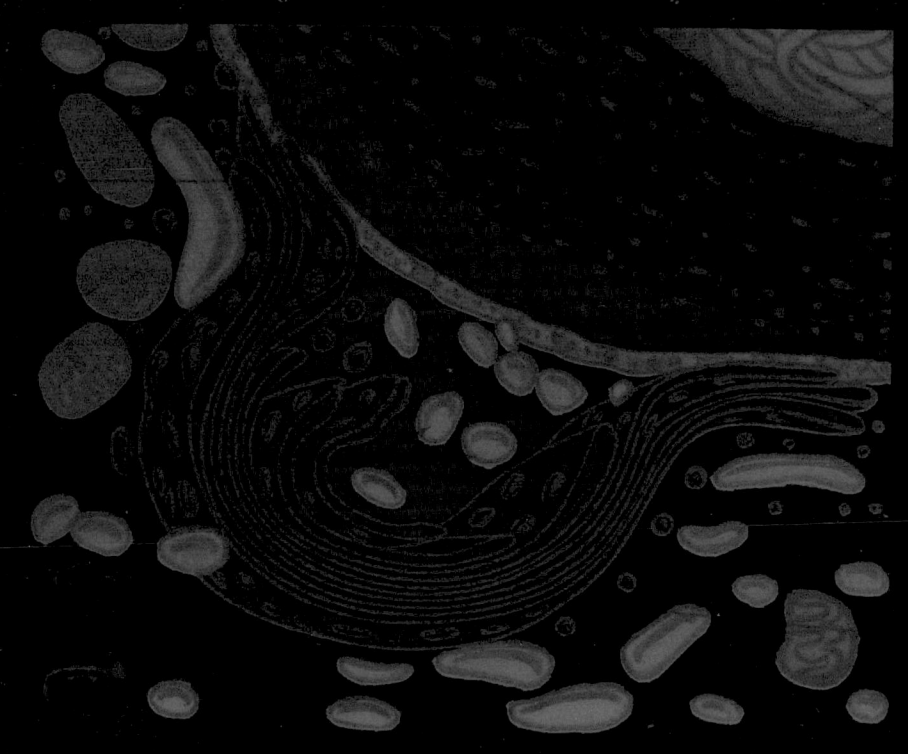